HOT TIPS FOR A COOL PLANET

CLARE EASTLAND

SOUTHGATE PUBLISHERS

First published 2007 by Southgate Publishers Ltd

Southgate Publishers Ltd, The Square, Sandford,
Crediton, Devon EX17 4LW

Printed and bound In Great Britain by
HSW Print, Rhondda

British Library Cataloguing in Publication Data

A CIP catalogue record for this book is available from the British Library

ISBN 1 85741 136 6

ISBN 9 781857 411362

Acknowledgments

- Special thanks to Chris Wakefield who designed and illustrated this book and whose passion for, and knowledge of, this topic proved invaluable.
- Thanks to Christine Mclennan and staff at the Centre for Alternative Technology and to Alan Dyer, Centre for Sustainable Futures, for their comments and input
- Thanks also to the following for checking through the manuscript: Christina Dymond, Denise Ross, Linda Lever.

Publisher's Note

It's impossible to produce a book of this size that cover everything; there are bound to be omissions. It's a fast-changing and hugely controversial topic but we have made every effort to make the content as accurate as we can. Most of the content is commonsense and it is hoped that families will use Hot Tips for a Cool Planet, and will at least make small changes to their lifestyle that will make a difference.

Contents

Introduction

It's all about the choices we make...

Why should we play our part?

- We can save money
- We can improve our health and that of our children and grandchildren
- We can improve the lives of people around us and some who are far away
- We can have lots of fun doing it
- We'll feel good and our children will enjoy long and happy lives!

We're now all aware of the major issue facing the planet in the 21st century – global warming. The question is – what can we do about it?

And can our individual actions make a difference?

The answers are – yes, we can make a difference, by doing lots of small things.

Governments all over the world are good at debating what to do, but individuals can act now and make a real difference – and by helping the planet we will also help ourselves, our children and our grandchildren. Green choices are not only about stopping global warming, they are also about enjoying life, having healthy products for our children and leaving the world in a better state for our grandchildren.

We do have the power to make changes and to persuade others over the world to follow suit. We can make a real difference by acting individually, and together. Despite the rapid industrialisation in countries like China and India, there is a growing movement against global warming and in favour of using the latest technology to reduce greenhouse gas emissions and pollution. We not only have the power to make a difference by persuasion but by collective action and by using our 'consumer muscle'. Once large companies realise they have to take note of consumer 'green' choices in order to sell more goods and make better profits, they will do so. What's more, many of the 'green' choices mean cost savings for everyone – businesses and consumers. We could all be better off!

We hope this book will enable you to take steps in making your family more sustainable, (see page 5, The Green Man Explains), to inspire you to make some changes to your lives, to make greener choices, and along the way you should have some fun together and save some money!

Start with the things which are easy for you – it doesn't have to be uncomfortable!

The GREEN MAN Explains

Climate change is beginning to be a real problem for us. There is increased flooding, drought in Africa, extreme events such as worse hurricanes, and it is beginning to affect food production. Global warming is also causing glaciers to retreat and the ice caps to melt, endangering animals such as polar bears. If the trend continues, sea levels will rise, flooding low-lying coastal areas in Britain, such as East Anglia, and entire island countries such as he Maldives.

There is a lot of evidence to show that the Earth is getting warmer. 2007 is predicted to be the warmest year on ecord, globally. There is also broad scientific agreement hat this is not just a natural phenomenon, but has been happening much faster since the industrial revolution and even faster in the last 25 years. And this warming of he Earth is changing climate patterns.

Climate change and global warming are the direct result of the way we consume energy. When we burn fossil uels, gases are released into the atmosphere. They include carbon dioxide (CO_2), methane, hlorofluorocarbons (CFCs), water vapour and a variety of other gases. They trap the sun's heat in the atmosphere and cause a gradual warming of the Earth. They are called 'greenhouse gases' because the effect is rather like the warming of a greenhouse.

By saving energy we can save fossil fuels and cut the amount of greenhouse gases released into the atmosphere (also called 'carbon emissions').

Human and technological solutions

Many of the solutions to this problem are already within reach or actually available.

The human solution

By just switching off the TV at night at the plug we can save energy at no cost. One-quarter of the energy consumed each year by the TV is used when it's turned off, but on standby.

The technological solution

Today most of us have mobile phones and other highly sophisticated technology but we're still using the same light bulb Joseph Swan invented in 1878. To save energy we could be using one that lasts ten times longer, uses one-fifth of the energy, and produces more light per watt – the energy-efficient light bulb.

If every household in the UK replaced one bulb with a compact fluorescent, we would save more than 330,000 tonnes of CO_2 from being emitted in just a year – that's equivalent to taking 77,000 cars off the road for the year.

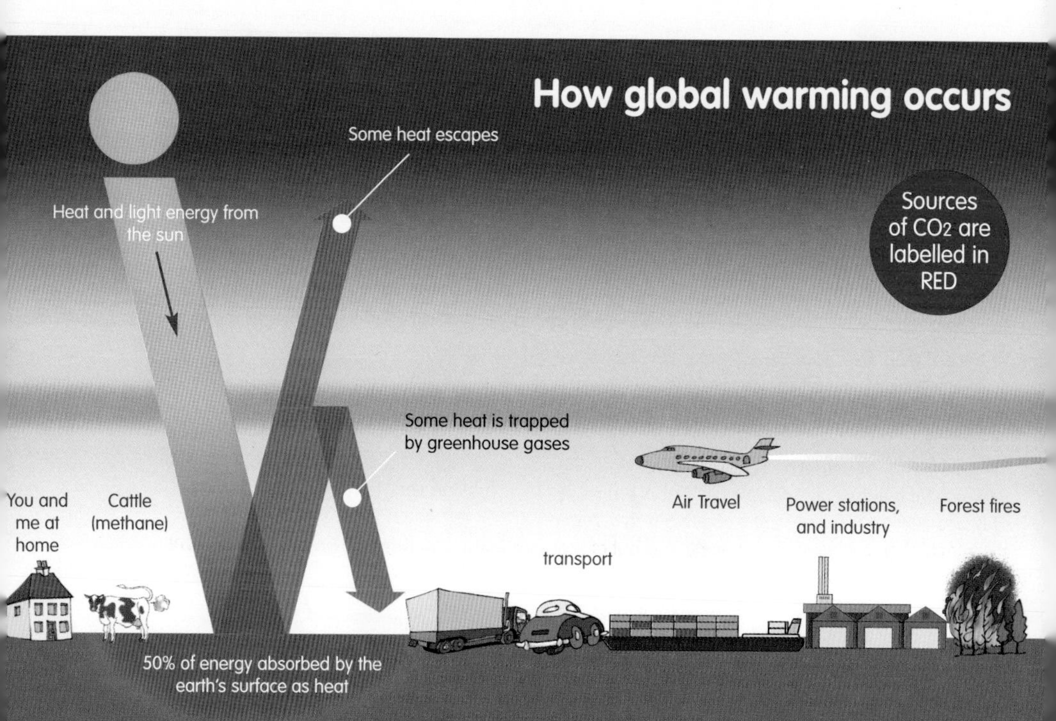

our carbon Footprint

No-cost savers

- Switch to a green energy supplier, which will supply electricity from renewable sources (e.g. wind and hydroelectric power) – this will reduce your carbon footprint. See page 15 and www.nef.org.uk or www.resurgence.org for information about renewable energy companies.
- Turn off electricity at the plug when not in use – (lights, television, DVD player, Hi Fi, computer, mobile phone charger, etc.)
- Turn down the central heating slightly (try just one or two degrees C).
- Turn down the water heating setting and check the central heating timer. Don't heat the house when no one is there.
- Always use your dish washer and washing machine with a full load and use off-peak electricity. Hang out the washing to dry rather than tumble drying it.
- Fill the kettle with just enough water.
- Do your weekly shopping on foot or have it delivered.
- Exercise outside on foot or bike rather than drive to a gym.
- Check your car tyres - soft tyres can increase fuel consumption by 8%.

Our carbon footprint is the impact our daily life has on the planet in terms of the amount of greenhouse gases produced, measured in units of carbon dioxide (CO_2).

The main way in which we contribute to greenhouse gases is through our fuel consumption, especially heating our homes, and how we travel. Electricity from fossil fuels is one of the biggest producers of carbon emissions, so every time we make a coffee or turn the television on we are adding to global warming.

It is better to reduce than to offset.

OFFSET...

Plant a tree

The average tree absorbs one tonne of CO_2 over its life time.

Low-cost savers

- Fit energy-saving light bulbs.
- Fit thermostatic valves on your radiators.
- Insulate your hot water tank, loft and walls.

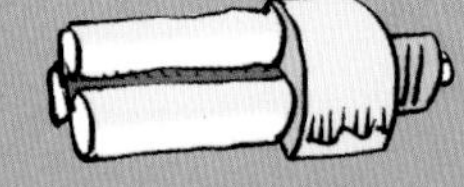

Longer-term savers

- Replace your old fridge / freezer (if over 15 years old), with a new energy-efficient one, rated A or A*.
- Replace your old boiler with an energy-efficient condensing boiler.
- Travel less and travel differently.
- Cut your air miles by holidaying in the UK or pay to offset them – it costs less than you think. Visit www.climatecare.org for precise calculations of flights and the cost of offsetting them.

See also pages 12-15 and 22-25.

Your secondary footprint

As well as your primary carbon footprint, you also make a secondary footprint through your shopping. For example, if you buy foods out of season, these have been flown or shipped in from other countries – all adding to your carbon footprint. Your secondary footprint is made not only by transporting things, but by the fossil fuels used to make the things you buy.

- Buy local fruit and vegetables, or grow your own.
- Try to buy clothes and other products from closer to home (avoid items that are made in the far-off places such as Australia, China and India).

See also pages 18-21, 30-31.

Activities

Calculate your carbon footprint

See www.carbonfootprint.com

www.co2balance.com or

www.bestfootforward.com

or simply type 'carbon calculator' into Google – there are many different calculators.

Your secondary carbon footprint

...is more difficult to calculate but take a look at your shopping habits and find out where things come from before you buy. See also pages 18-21 and 30-31 (food and shopping).

The Challenge

The UK has a target to cut its carbon emissions by 60% by 2050. Can you reduce your own carbon footprint by at least 10% each year?

Our carbon footprint is the amount of greenhouse gas emissions we are responsible for. It has a direct impact on global warming and climate change. It is best to reduce our carbon footprint if we can – and most of us can to some extent. It's difficult to live a 'carbon neutral' life in our society, but some people are trying – notably Ashton Hayes (visit www.goingcarbonneutral.co.uk).

Perhaps we should also think about carbon offsets – balancing our own carbon emissions by reducing carbon emissions elsewhere in the world or by doing something which takes carbon dioxide out of the atmosphere. By planting a tree you enable carbon dioxide to be absorbed from the atmosphere. This process is called 'carbon sequestration'.

You can buy carbon dioxide credits and then not use them. This 'locks up' CO_2 and stops other people from producing emissions. You can also invest in companies and organisations who research and develop renewable and sustainable technologies. These might include renewable energy, biomass fuels (plant fuels burnt to produce energy), recycling, energy-efficient vehicles. And you can buy energy-efficient technologies, such as solar panels, and give them to organisations in developing countries. However this is a complicated area. To find out more visit
www.carbonplanet.com
www.carbonneutral.com
www.foe.co.uk or
www.CRed.uk.org

Shocking statistic

- The average UK household emits 6 tonnes of CO_2 a year.
- The average UK person has a carbon footprint of 9 tonnes of CO_2 a year including secondary emissions – that's equivalent to 5 hot air balloons full.

reclaiming Waste

Remember the Rs!

Reduce your waste

One important and easy way to help the planet is to think about what we use, and try to reduce our impact. We can try to use fewer of the world's resources, and throw less stuff away. This can be 'win-win', spending less money on things we don't need. For example, with very little effort we can throw away less packaging.

RESOURCES
use fewer of them

REDUCE
your waste

RESCUE
things which can be re-used or recycled

REFUSE
packaging or goods which use lots of energy

REUSE & REPAIR
whenever possible

RECYCLE & RECOVER
resources

RESIST
the temptation to buy

Shocking statistic

- Each household in the UK throws away around a tonne of waste each year.

Rescue and recycle the following treasures:

Find out where these items can be recycled, either in the kerbside collections or at local recycling banks.

Refuse the packaging!

- When shopping always say 'No!' to additional bags, boxes etc.
- Buy 'loose' products such as vegetables.
- Buy items which can be refilled, e.g. washing-up liquid.

The Challenge

Reduce the amount you throw away each week. Can you cut it by half? If you normally put out four bags can you cut it to two?

If everyone managed to do this we'd save about 15 million tonnes of landfill a year.

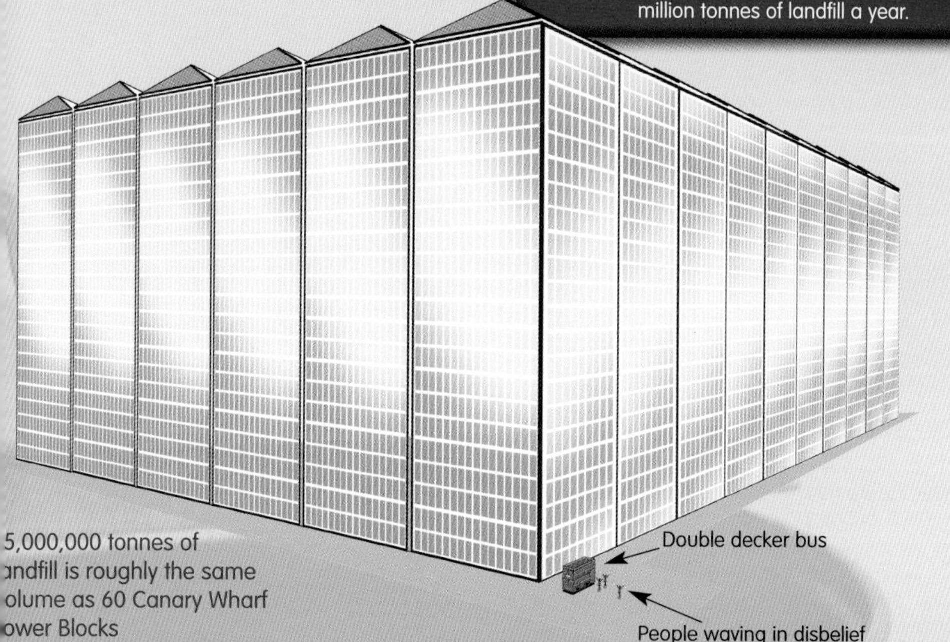

5,000,000 tonnes of andfill is roughly the same olume as 60 Canary Wharf ower Blocks

Reuse and repair

Reuse plastic carrier bags or better still buy a bag or basket to use all the time.

Use rechargeable batteries.

Buy reusable cotton nappies – disposable nappies make up about 3% of household landfill waste.

Buy second-hand furniture.

When you've finished with something sell it (e-bay, car-boot sales) or give it away to charity shops, magazines to dentist's waiting rooms, or visit www.freecycle.org There are freecycle groups all over the UK which enable you to give away things easily – 'buyer' collects.

Repair electrical goods and furniture if possible. If not these too can be recycled – check the phone book or the web for details of local organisations.

Recycle

ecycle materials when you've exhausted the other ossibilities. See the box on page 8 for the easiest things recycle. Everything except the plant remains can be taken to your local household recycling point, or put in a kerbside recycling collection. Check what your local authority can take.

Plant remains, kitchen waste, (but not meat), and garden waste can be recycled at home in a compost bin, and used to improve the garden.

Activities

1. Take a look in the bin!

Safety warning: beware of any sharp metal or broken glass etc.

A messy activity the children should enjoy! Before you put out the rubbish one day, go outside, put on plastic gloves or plastic bags over your hands and just see what's in there. What valuable resources are you going to bury in landfill? Tip the bin out onto a sheet of plastic or old newspapers or pick it over, sorting into separate plastic bags.

Is there anything that you or someone else might be able to reuse? Is there anything that could be repaired? Did you really need to throw this away? What could be composted?

2. Check the shopping basket for packaging

You can either do this as you go round the supermarket or perhaps the first time when you get home. Check what you've bought to make sure you haven't got more packaging than you need – avoid things which are wrapped twice or which use heavy-duty plastic packaging. Try to buy loose (it's often cheaper too) and go for paper packaging or glass, which are easily recycled, rather than plastic.

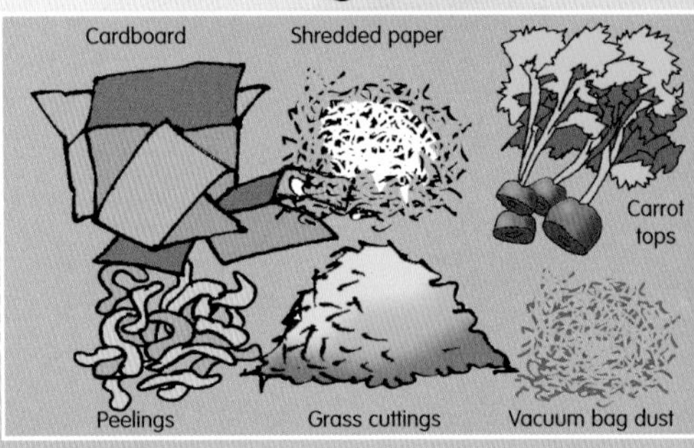

Easy compost bins...

Purpose-made bin from a garden centre. You may get a subsidy from your local authority

DIY bin from 3 wood palettes, with 1/2″ wire mesh front, and carpet cover

DIY bin from a stack of old tyres

Shocking statistics

- Each year we throw away 17 billion plastic carrier bags – 290 bags for every person in the UK.
- Each household throws away over 200 kg of paper and card in a year.

Old plastic bin with base cut away

Compost pen using 1/2″ mesh around four stakes

3. Making compost

You will need some kind of bin in the garden to make compost in. You can buy one – try your local authority for a subsidised one or a local garden centre or look on the web. Or you can easily make one – see the pictures on page 10.

Build up the bin gradually with organic waste which will rot fairly quickly: vegetable and fruit peelings, grass mowings, weeds, tea leaves, vacuum dust and small quantities of non-glazed cardboard. Avoid meat, bones, large twigs, diseased plants and perennial weeds.

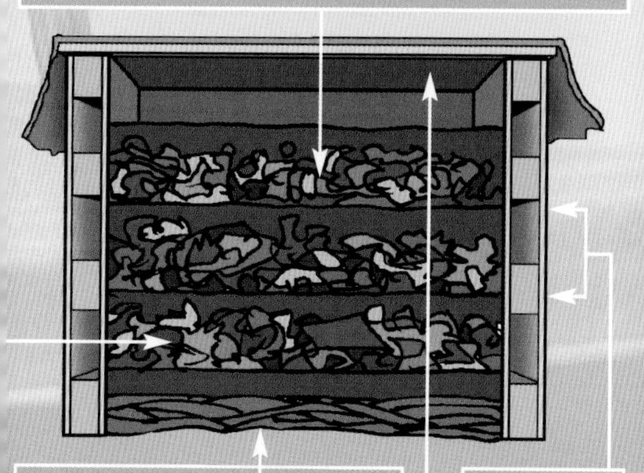

Larger sticks or hedge trimmings on the bare earth. This allows water to flow away, air to circulate and worms and other creatures to find their way in.

Build up in layers with a little soil over each 15cm layer.

Cover with plastic or an old carpet to keep in heat and stop it getting too wet.

After a few weeks mix it all up to help it rot or turn it into the next bay if you have space for two. Once it looks like soil all through you can spread it on the garden as a fertiliser.

Good compost is produced with warmth, moisture (not too wet) and a good balance of materials, mixed well and not in large lumps. The best compost I've ever had was made in an old coal store, with open sides and an old window over the top to act as a solar heater. Experiment and see what works for you.

See: www.gardenorganic.org.uk/organicgardening/compost and www.envocare.co.uk/makingcompost

The GREEN MAN Explains

Waste is what we throw 'away'. Everyone and everything we do makes waste. Waste is important for a number of different reasons.

Firstly there is no 'away' in 'throw away'! Everything we dispose of has to go somewhere else. So what do we do with the waste we produce? Most of it (72%) goes into landfill. Many landfill sites are nearly full and we are running out of places to put it. Landfill is a costly option and not something any of us wants as a neighbour! Landfill has an impact on the environment. As the waste breaks down, mostly very slowly, dangerous chemicals and metals are produced which have to be dealt with. Methane gas, which is a greenhouse gas (affecting global warming, see page 5) is often produced. 9% of waste is incinerated, which at least means we can recover some energy from it, but still results in CO_2 emissions and contributes to global warming.

Secondly when we throw something away we lose the natural resources, energy and human resources used to make it. Most of the resources we use in this way are finite – they cannot be replaced and will eventually run out.

When we throw something away we can no longer use it as a resource. For example, if we throw glass bottles away instead of recycling them they take thousands of years to break down in the soil. When we next want a glass container for our liquids we have to make glass from scratch. This puts unnecessary pressure on the world's resources and means less is left for future generations. Extracting the raw materials, manufacturing and distributing the glass all take additional energy. Recycling two glass bottles saves enough energy to boil a kettle of water. On average we each throw away 131 bottles a year; multiply that by the number of people in the UK and that's a big saving on energy and carbon emissions. See pages 6-7 and 12-15.

extravagant Energy

We all waste energy around the home. As a nation we waste about £5 billion of energy each year. The great thing about saving energy is that it's mostly a way of saving money too. There are many things we can do, some of which cost nothing except a little thought and which save us money. Others cost a little more, but also save more in the long term.

No-cost savers

- **Turn the heating down**, wear a jumper and thicker socks or slippers. If we turn the main thermostat down by 1°C we can cut our heating bills by up to 10% per year. Only heat rooms you are using at the time.
  If you have heating on a timer check the clock times, turn down bedroom radiator thermostats and use a thicker duvet. A cooler house is better for you – you're less likely to catch colds and other nasties!
- **Electricals on stand-by?** Turn off the TV, video, computer, mobile phone charger, etc. when not in use. You are probably spending £1 a week on standby power.
- Choose the right size pan for whatever you're cooking, use as little water as possible and put a lid on it. Cut food up smaller and it will cook quicker.
- Turn off lights in unoccupied rooms.
- Draw the curtains at night.
- Only fill the kettle with the amount of water you are going to use now.
- Use full loads in the washing machine and the low temperature wash – 40°C or less and use off-peak electricity.
- Use a washing line instead of a tumble drier when the weather is fine.
- Have a shower instead of a bath (except if it's a power shower) and save water too – see page 16).
- Don't buy or use home air-conditioning – this uses huge amounts of electricity. Instead, open the windows, close the curtains, and if really necessary use a fan. Or investigate 'passive cooling' – see www.carbontrust.co.uk
- Move furniture away from radiators so that the heat can radiate into the room.

Shocking statistic

- The amount of heat lost annually through roofs and walls in the UK is enough to heat three million homes for a year.

Low-cost savers

- Low energy and long life light bulbs – replace old bulbs (save £7 a year each).
- Insulation – put a thick jacket around the hot water tank (save £15 a year); insulate water pipes (save £5 a year).
- Insulate the loft with 270mm of insulation (save £70 a year).
- Draught-proof around windows and doors using DIY strips (save £10 a year).
- Buy energy-efficient electrical appliances (an A or A*-rated fridge freezer can save £45 a year compared with an old one).
- Invest in solar-powered or clockwork gadgets – radio, mobile phone charger, garden lights – see www.cat.org.uk
- Buy a plug-in powermeter so you can see how much electricity each appliance is using, or a device which monitors your household use. Visit www.cat.org.uk

The challenge

Can you save 20% of your energy consumption? The Energy Saving Trust is asking everyone in the UK to save 20% of their energy. Visit their website www.energysavingtrust.org.uk/commit.

Longer-term projects

Count up the cost-benefit – find out how much the work will cost, divide that by the fuel savings each year and you know the number of years it takes to break even. After that you're saving.

- Cavity wall insulation – this can cost anything from about £300 but gives a saving of between £60 and £160 a year. Grants are now available so it may pay back sooner than you think – check with your energy supplier or local authority.
- Thick carpets with underlay, woodblocks or cork tiles can save about 10% of heat compared with bare floorboards.
- Double-glazing can save about £20-30 a year.
- Replace your old boiler, fridge or washing machine with a new one and you might save more than 25% on your bills. Install a condensing boiler and you could save 10% of your energy use.
- Check out solar panels – at the moment they take some years to break even, but they are getting cheaper and more efficient and you may be able to sell electricity back to the utility company – see www.est.org.uk

Activities

1. Read your electricity meter

Work out how much electricity you are using. Most meters have digital read outs, but some older ones have dials. A dial meter has six dials. You need to read the first 5 dials, starting on the left (the 6th dial is for testing only). If the pointer is between two numbers always read the lower number. If you're not sure if a pointer has passed a number check the next dial to see if it is approaching the 0 or has gone past it.

Each week read your meter. Take last week's reading from this week's and you'll have the number of kWhs (Kilowatt hours) used. A kWh is a 'unit' of electricity and you pay according to how many of these you use. You could keep a diary or family chart to show each week's usage.

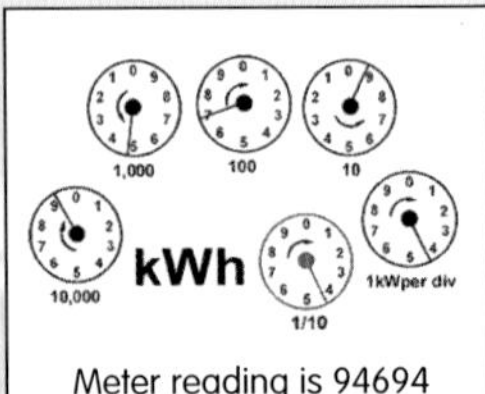

Meter reading is 94694

2. Energy use

Work out how much you use each week – read the meters (electricity and gas) at the same time each week and take the lower figure from the higher one. Look at fuel bills such as gas, oil, coal and wood to find out how much you use and how much it costs. Check your electricity bills and find out how much you pay for a unit (kWh) of electricity. Then you can monitor your use and see what you're saving.

3. Home energy audit

Make a list of all the things in your home that use electricity or other energy. For each one think about how it is used and whether you could reduce the energy consumed without discomfort. Check the heating system, hot water system, lighting, appliances and insulation against the savers on pages 12 and 13.

Buy a simple plug-in powermeter which will show you how much each electrical appliance uses.

4. Enjoy the savings!

Once you've done your audit, work out how you can save energy over the next few months and make sure everyone in the household knows and agrees. You could try having one person responsible each week for checking up and switching off, and keeping everyone on track – the youngest members of the household often enjoy this! Then compare the next fuel and electric bills with the last ones you had or compare with the same time the year before to account for the seasons. Work out what you've saved as a family and decide together how you will spend it – perhaps a day out somewhere? Or you could invest it in more energy-saving and see how much more you can save!

5. Make a haybox cooker

Hayboxes are insulated boxes for slow cooking. They are good for soup, stews, brown rice or anything else that needs slow cooking. You will need a cardboard box, insulating material and a pot with a well-fitting lid for the food, such as a casserole. You may like to experiment with different kinds of insulating material. Originally hay was used, but polystyrene and crumpled newspaper are fine. If possible make the box to fit a particular pot, as the less air round the sides of the box the better. But allow an inch or two for insulating material between the pot and the box, all round, underneath and above.

To use a haybox you need to bring the contents of the pot to the boil on an ordinary stove or in a microwave first. Then put the pot in the haybox and close the lid of the haybox quickly. Experiment with cooking times as this

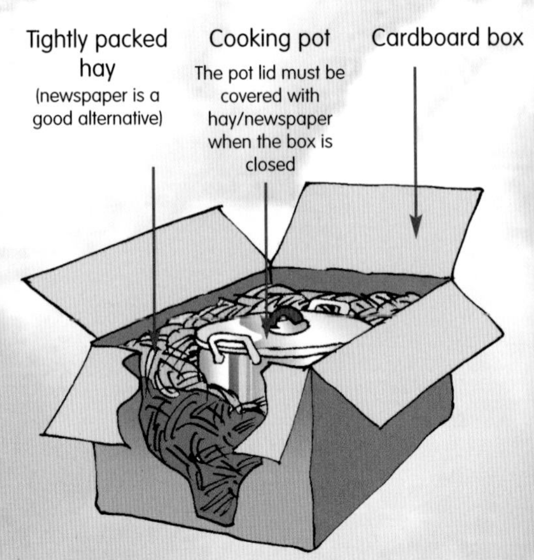

Haybox cooker

will vary from box to box. Try not to remove the box too often. Each time you do remove it you should bring it to he boil again on a stove before putting it back in the box. Approximate cooking times are: stews 3-5 hours, entils 1-3 hours, milk puddings 1 hour.

6. Experiment - make a solar water heater

Experiment with making a small solar water heater. Use an aluminium pie-plate or ready-meal container. If possible compare two – one which is wide and shallow and one which is smaller and deeper, to see which works best. Make sure they are empty, clean and dry. Then paint the inside of each with non-water-soluble black paint. When the paint is dry add about 2 ablespoons of water to each one. Place the heaters on op of a stack of newspapers in the sun or 10 minutes. Use a hermometer or a finger o test the temperature. Which works best? What happens if you leave hem in the sun longer?

ou could try comparing ther colours of eaters by seeing which can elp to melt n ice-cube uickest.

How much energy do I use? (Amounts used shown in **Kilowatts**)

Energy saving light bulb	0.007-0.015 kW
Fluorescent light tube	0.04-0.1 kW
Tungsten light bulb	0.06 kW
Hi-fi	0.06 kW
TV	0.08 kW
TV (big screen)	up to 0.5 kW
TV on stand-by	up to 0.02 kW
Hairdryer	1 kW
Electric kettle	2-2.5 kW
Electric cooker	12-15 kW

You can check the power rating of your own appliances by looking at the backplates to see the number of kW. If the power rating is shown in W (Watts), just divide by 1000 to get the kW.

Most energy in the UK comes from burning fossil fuels – coal, oil and gas. There are only finite amounts of these in the world; they are non-renewable. As fossil fuels are finite, this is not a sustainable way to live – we won't leave enough fossil fuel resources for our children and grandchildren. And fossil fuels are needed to make many important things such as plastics, medicines and chemicals.

Burning fossil fuels creates carbon-dioxide (CO_2) emissions, (see page 5). We need to reduce the amount of energy we use in order to reduce the amount of fossil fuels used up and the amount of CO_2 which goes into the atmosphere.

Nuclear power does not produce CO_2 emissions and so contributes less to global warming than burning fossil fuels, but it does require uranium which is a finite resource, and also uses energy in the mining and processing. Other hazardous environmental effects and potential security problems also have to be considered.

Green energy is generated by sun, wind or water power and does not use fossil fuels. It is also carbon-neutral (except for any carbon used to make the actual equipment) and does not release CO_2 into the atmosphere, so it does not contribute to greenhouse gases, global warming or climate change. Most of the main utility companies sell green energy, but it sometimes costs more than conventional energy, and it is difficult to know for sure how much difference you are making, so check at www.greenelectricity.org Or you could switch to a green electricity supplier, such as Good Energy or Ecotricity. Alternatively you could create your own green energy by installing solar panels – visit www.cat.org.uk

Shocking statistic

- If every household installed three energy-saving light bulbs, the energy saved in a year would supply all street lighting in the UK .

wonderful Water

Low-cost savers

- Fix any leaks or dripping taps.
- Fit a 'hippo' in the cistern. A 'hippo' is a bag which takes up space and saves water when you flush the toilet – 'hippos' are available from the water companies.
- Have a jug water filter and use it to refill plastic water bottles to take out. Don't buy expensive bottled water which has lower health standards than tap water.
- Collect rainwater in water-butts and use a watering can. If you must use a hose, use a trigger nozzle to control the water flow. Don't use a sprinkler.
- Use collected rainwater to wash the car, rinse vegetables or anything else which doesn't really need tap water.

Water is essential for all of us and we are lucky to have safe drinkable water available at the turn of a tap. Many millions of people don't!

Consider fitting a water meter. It will raise your awareness of how much water you use and encourage you to avoid wasting water. With a meter, by saving water you also save money on your water bills. You could save 5-15% in household water use and about the same in water bills. Of course if you save on hot water you'll save on energy too.

No-cost savers

- Have a 5 minute shower instead of a bath.
- Don't leave the tap running when you clean your teeth.
- Put a plug in the bowl to wash your hands.
- Wash vegetables or fruit in a bowl of water. Use the left-over water for watering garden or house plants.
- Use as little water as possible when you boil water in saucepans and boil the kettle.
- Always use full loads in the washing machine and dishwasher (half-loads use almost as much water and electricity as a full load).

How much water do we use? (Volumes used shown in litres)

Activity	Litres
A full bath	75-100
Normal shower (5 mins)	5-7
Powershower (5 minutes)	85
Wash hands and face	3-9
Flushing toilet once	10
Flushing toilet once with 'hippo'	7
Brushing teeth, tap running	5-15
Brushing teeth (tap on and off)	2
Cooking and drinking etc per day one person	10
Washing up in the sink	6-8
Dishwasher	30-50
Washing machine one load	120

The challenge!

Can you save on your water consumption by 20%? The average person in the UK uses about 150 litres each day. The average person in the developing world uses 10 litres of water a day. To save 20% you need to reduce your consumption by 30 litres.

Activities

1. Home water audit

Make a list of all the things in your home that use water. Check the box below left.

Do you use water in the garden (hosepipe 550-1000 litres an hour) or for cleaning the car (80 litres)?

Count up the number of times each one is used each day. Then work out roughly how much water you use in a day. How you could reduce the amount of water you use? Or visit

www.bristol-water.co.uk/environment/waterSavingAudit.asp and try their water audit.

2. Make your own 'hippo'

You'll need an old plastic water bottle. Take the lid off the cistern and check that the bottle will fit inside. Fill it with water and place it inside the cistern. Check that the cistern still flushes properly and then put the lid on. You might like to decorate your bottle with eyes, mouth etc to make it look like a real hippo!

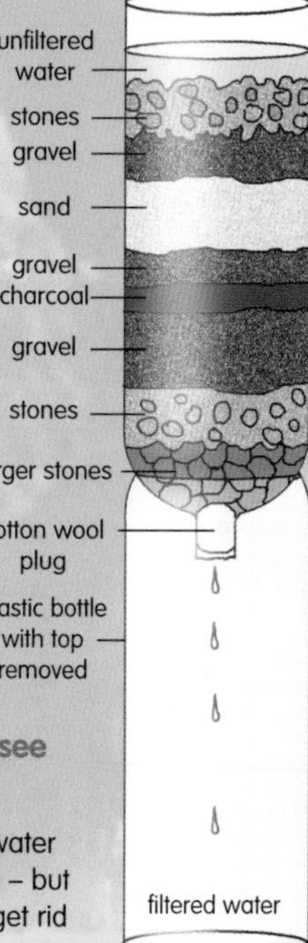

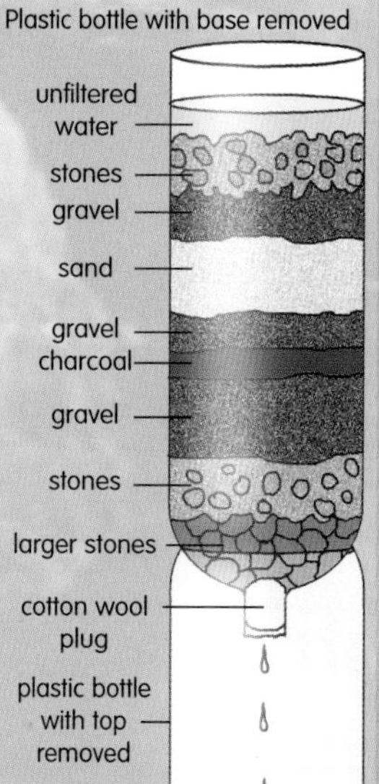

3. Make a water filter (see illustration right)

Try using the filter to make dirty water clean enough to use for washing – but don't drink it! – the filter doesn't get rid of germs.

The GREEN MAN Explains

Wondering why we should save water? Well, as the climate gets warmer (see page 5) the south and east of Britain is likely to experience more summer droughts, and the north-west may be wetter. Britain will probably have about the same amount of rainfall, but in bigger doses, so it's really important that we collect what we can when we get it. Our population is increasing and our individual water use is also increasing so we'll need more water in the future.

The water we use has to be cleaned and this takes energy, so contributing to global warming. So by saving water we're also saving energy, and reducing CO_2 emissions. By saving on hot water we're saving even more energy of course.

Water treatment

The water we drink has been through various stages. Water is taken from rivers and reservoirs, passed through screens to exclude solid material, and pumped to the treatment works.

Ozone and ferric are added to the water to make small particles which attract impurities. The water is slowly stirred so that the particles grow in size and are easier to remove when passed through a clarifier and sand filters. The filtered water is collected in a large tank and disinfected with ozone and granular activated carbon. Finally chlorine is added to kill any remaining micro-organisms such as bacteria before the water is stored and pumped to your tap.

visit www.wateraid.org.uk/uk/learn_zone/games to learn more about water and play the games.

Shocking statistics

- We use 70% more water in the UK than we did 30 years ago.
- Only 1% of the water on our planet is drinkable. 3% is fresh water and 97% is sea water, but most of the fresh water on the planet is locked up in the ice sheets at the North and South poles.
- Over 1 billion people do not have safe drinking water – that's one in six of the world's population.

fabulous

The choices we make about what we eat can have a big impact on the planet. Food is the area in which it's easiest to make a difference. Sometimes the choices can be difficult to weigh up, (imported field-grown Spanish tomatoes or British ones grown in heated greenhouses? – in this case it's more energy efficient to buy the Spanish ones!). But it's not about getting every choice right – just the general trend. And what and how we buy will influence the shops and food companies too.

No-cost savers

OFF

- Buy what's grown locally and in season – avoid energy used in **transportation** (beans flown from Kenya, 3,600 miles) or **growing** (strawberries grown in heated greenhouses).
- Buy from farmers' markets or farm shops, or small shops which source their produce locally. The food is usually sold loose and you know it's been grown to strict UK laws on pesticide use, etc. And it's usually produced for flavour and quality – it hasn't been produced to be uniform and survive long journeys.
- Join a local organic box scheme, but check to see where the produce comes from. See www.vegboxschemes.co.uk
- Avoid buying more food than you need, and use up left-overs in soups or stir-fries.
- Avoid excess packaging – cut down on waste and energy use.
- Buy fresh food rather than processed food – better for your health (less processed fats, salt and sugar) and less energy has been used to produce the food.
- Eat less meat – the biggest global source of CO_2 emissions is in the production of animal feeds, housing and transport for large scale livestock production. Animals themselves produce methane, another greenhouse gas. Meat takes a lot of energy to produce so buy smaller amounts of good quality, locally grown meat and don't waste any of it!
- Avoid ready meals – they're twice-cooked, which is a waste of energy, the ingredients are usually the cheapest available (which means intensive farming using chemicals which pollute the soil and rivers), and they always have plastic and card packaging.

The challenge!

Have a blind taste challenge, comparing organic produce with similar non-organic. Try something raw like tomatoes or apples and also something cooked such as your own home-made dish compared with a ready-meal. Which foods win the challenge? Can you cook a meal for the family for less than the price of the similar ready-meals?

Shocking statistics

- 95% of our fruit and 50% of our vegetables are imported.
- 25% of all HGVs on our roads are carrying food.
- 25% of our national energy bill is used in growing, processing and transporting food.

Why you want to know what's in what you eat!

- **Family health** – you can't see the chemicals which have been used to produce your food, and they build up slowly in the body. Those really perfect looking apples or cabbages may well have been drenched in pesticides, whereas if you find a slug you know it won't poison you! Fresh foods have few or no additives. If something has been designed to keep for a long time then it probably contains lots of additives.
- **Pesticides and pollution** – organic produce has been grown without dangerous chemical pesticides and fertilisers so has a less polluting effect on the environment. And if you cook food from fresh ingredients you avoid the food factory adding to greenhouse gases.
- **Animal welfare** – buy free-range or organic, especially when it comes to eggs or chickens. Non-organic and non-free-range chickens are usually kept in huge barns with no access to the outside and are cheaply produced by over-feeding and the use of growth hormones and antibiotics.
- **Fairtrade** – for coffee, tea, bananas, chocolate and similar goods which we can't grow locally, look for the Fairtrade logo to be sure that small farmers in developing countries have been paid fairly for their products, and plantation workers get a fair wage. Small producers tend to farm less intensively than big ones, too, so they're better for the environment. Find out about Fairtrade towns too at www.fairtrade.co.uk.

- Buy local
- Buy seasonal
- Avoid packaging
- Buy fresh

An average apple is sprayed with 36 chemicals in 16 sprayings

What does organic mean?

Organic farming does not use dangerous chemical pesticides or fertilisers. This means that you don't need to peel carrots, just wash them. Organic farms are better for wildlife and the environment. Many people say organic food tastes better and is healthier (organic milk is higher in Omega 3 fatty acids – the ones that are good for you!), but make sure it's fresh – the shorter the time between field and plate the tastier it will be, whether it's organic or not.

Animals kept organically are reared with no artificial hormones, pesticides or antibiotics, and are kept under more natural conditions.

The Soil Association mark guarantees that a product is organic. Visit www.soilassociation.org.uk to find out more.

Soil Association
the heart of organic food & farming

Activities

1. Food miles

Look at the ingredients for one meal or one shopping basket/trolley. Check the country of origin of each and then use a world map to work out the number of miles (minimum) the food must have travelled to get to you. The country of origin is usually displayed on the shelf or on the packet or label. You could even draw a map with lines connecting the UK to each of the other countries to show where your food has come from.

Apart from the distance from country of origin it's difficult to find out how far food has travelled and by what transport. This can be important environmentally, as a long journey by ship, for example, has less impact than a shorter one by air or road.

You can have a direct effect on food miles by going food shopping on foot. But if transporting 10 heavy carrier bags on the bus by yourself seems impossible, at least you can make good choices in the shop!

2. Read the packet!

Next time you buy a ready-meal look at the ingredients in small print on the packet. Do you know what they all are? They may be safe, but are they good for you? Try a web-search to find out what they are and why they're there. The main ones to look out for and avoid are salt, fats, especially trans-fats, and monosodium glutamate (MSG) – also known as E621. Some supermarkets are even going to label their products to show their carbon footprint!

3. Grow your own food

– in a window box, patio tubs or a garden. Many crops, such as salad leaves including rocket, tomatoes, chillies, or dwarf beans will be happy in pots or tubs if they're watered. In even a small garden you can grow potatoes, carrots, onions, spinach, beans and fruit such as strawberries and raspberries.

If you don't have even a window box try growing mustard and cress on soft wet paper (traditionally blotting paper) or sprouting beans in a jar.

If you are really keen get an allotment – they're usually let out by the local council. You can rent a plot of land for about £30 a year and grow your own fruit and veg.

To grow food organically all you need is compost (see pages 10-11) or organic fertiliser. Visit Henry Doubleday Research Association (HDRA) website for lots of good advice: www.gardenorganic.org.uk and also see www.permaculture.org.uk

4. Collect free food in the countryside

Late summer or autumn are the best times for this. Pick blackberries (for blackberry and apple crumble and juice), nettles (for soup, in Springtime), dandelion leaves (blanched, for salad), crab apples for jam, and rose-hips to make syrup. Avoid picking from road-side hedgerows as these fruits may be polluted. You might also like to go mushroom-picking, but it's best to join a guided 'fungi foray' for this and learn from the experts.

5. Visit a farmer's market...

or farm shop, or city farm and have a look at the produce. Ask where it has come from and buy some to take home. Work out the food miles and compare that with something similar bought from the supermarket. Many farmers' markets will only sell food from a specific area around.

The GREEN MAN Explains...

Shocking statistic

- The UK food chain (agriculture, processing, transport, etc) is responsible for about 22% of UK greenhouse gas emissions – that's about the same as all road transport.

...how food choices affect the planet

What is sustainable food?

Our present food system is unsustainable. Sustainable food is good for people and the planet. It is...

- environmentally friendly (good for the planet and the local environment),
- socially just (fair for people all over the world) and
- economically useful. It provides jobs, is a good use of resources and produces healthy food.

At the moment we are 'eating oil'. Huge amounts of oil is used in making fertilisers, pesticides, etc, for use in farming, in growing food (heating, irrigation, farm machinery), in processing, packaging and transport. Many more calories of energy is used in growing the food than our bodies get out of it.

What are food miles?

Food miles are the distance a food travels from field to plate. This transport uses fuel and of course the CO_2 emissions add considerably to global warming. Road and air transport of food is worse than sea or rail but even they use energy. Agriculture and food now account for nearly 30% of goods transported on our roads. So imagine how much congestion could be cut back by reducing food miles.

Food travels further today partly because supermarkets have centralised systems of distribution. A kilo of apples or a pint of milk can be transported many miles from the farm to be packaged at a central depot and then sent many miles more to be sold. It may even eventually come back to near where it was produced. Also, ingredients used in food processing travel from factory to factory, before the final food is packaged and transported to the shops.

The amount of food we import has also risen enormously. Much of this food is flown in; for example strawberries and beans come from warmer places at all times of the year.

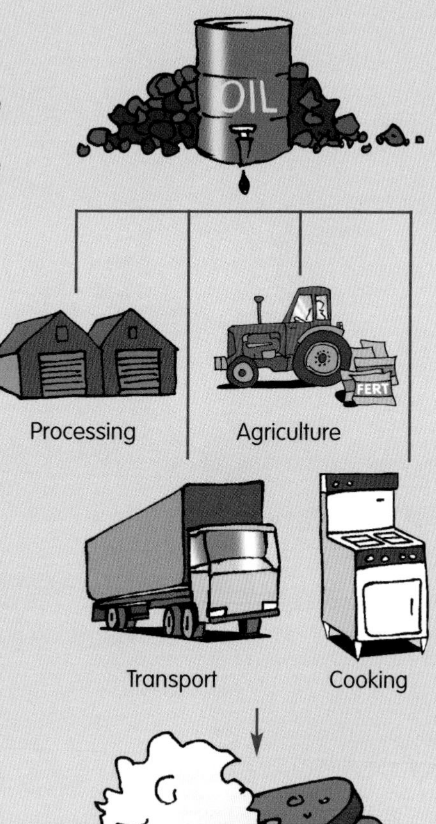

We can make a difference by making choices about the food we buy and eat. We can choose where we buy and what we buy. Sometimes it can be difficult to decide on the best option, but food labelling is gradually improving. We can look out for Fairtrade and Soil Association labels, for organic produce and things grown locally. We can avoid produce from far away, processed food and additives.

Terrible Transport

Of course if we all walked everywhere we'd cut our greenhouse gas emissions and the world would be a healthier place. But that's not going to happen and many of us need our cars for essential things, like all that shopping! However, we can still make healthier choices for ourselves and the planet at least part of the time.

Before you make a journey weigh up the options – is it really necessary to use the car this time? Could you improve your health or save money by walking, cycling or using public transport?

No-cost savers

- Journeys under 2 miles are big polluters and use more fuel. Why not walk or cycle?
- Would it be cheaper to take the bus or walk into town rather than pay for parking?
- Share your car with friends or neighbours – offer lifts or combine journeys.
- For long journeys check out low saver fares by coach or rail.
- Make walking or cycling part of your leisure activities and get fit! The more you do, the more you'll find you want to.
- If you are a two-car family, do you really need two cars, or could you manage with one? You could save a lot of money.

OFF

Shocking statistic

- 25% of UK greenhouse gas emissions are from transport.

Greener driving

Next time you buy a car check the different models and brands and buy for fuel economy and safety. It will save you money. Find out about diesel engines which can also use biodiesel fuel and hybrid cars.

Always turn the car engine off when you are stationary – idling causes lots of pollution.

Remove any roof or bicycle racks when not in use.

Avoid carrying heavy things you don't need.

Keep to the speed limit – driving at 50mph takes 25% less fuel than driving at 70mph.

Get your car serviced regularly and check your tyre pressures - soft tyres can cost up to 8% extra fuel.

Drive smoothly, avoid accelerating fast or braking hard – this can save you 30% on fuel.

Driving at more than 40mph with the window open costs more fuel than having the air-conditioning on.

Activities

1. The school run

Do the kids still go to school by car? If so, examine the options and talk to fellow-parents and teachers. Ask the kids to work out the most sustainable option and the best route. Can you save time and money?

Some schools organise 'walking buses' led by adults which pick children up along set routes and keep them safe. See www.travelwise.org.uk

If you can't walk everyday, could you manage it once a week?

Low-cost savers

- Buy a bike (preferably second-hand!). They are cheap to maintain and get you fit. Many areas now have good cycle routes away from traffic and they're free to park.
- Look at your journey to work - if it's too far to cycle could you use public transport? You'd have time to read instead of raging at the traffic! Or could you car share?
- Look at the children's journeys to school. Could they safely walk, cycle or use public transport with a rota of parents? Ask the school to provide safe, secure bike parking.
- Try a car-share club – see www.car-plus.org.uk to find out about car hire clubs and lift sharing.
- Don't use domestic flights – for journeys in the UK go by train or coach, or even by car!
- Try holidaying in the UK, or the 'new' slow travel – experiencing the journey as part of the holiday, by going by train or coach.

Shocking statistics

- A large 4x4 emits 50% more CO_2 than a small car.
- The average UK driver emits 4 tonnes of CO_2 a year. This is one medium sized hot air balloon full of CO_2.
- A holiday to Australia produces as much greenhouse gas emissions as three petrol cars in a year.

Walking to school – the plusses

- Save £300 a year on petrol, car maintenance, etc.
- Have quality time with your child.
- Teach road safety effectively.
- Meet other people in the community – you'll soon recognise others on a regular walk – say 'hello' and make friends.
- Improve the family's fitness and health.
- Reduce your child's risk of diseases like heart disease later in life.
- Develop healthy lifestyle habits.

2. Get on yer bike!

Help the children to investigate why it takes less energy to cycle the same distance as to walk it. How do the bicycle gears work? Could we use 'pedal power' for other things? Children may like to invent their own 'pedal-power' machine to do something else.

3. Plan a 'green' family holiday

which everyone will enjoy. Avoid air travel and if possible go by train or bus/coach. If you want to go abroad take the channel tunnel or the ferry to France or Spain. Think about cycling, walking or other activities which are carbon neutral, like sailing or canoeing. Think about where you can stay – have you tried camping, bunkhouses, youth hostels or eco-cabins?

The challenge

Can you cut down your car use by just one or two journeys a week?

Try that for a month and look at other possibilities which could be more fun than driving, and save you money!

Feeling guilty?

If you can't cut your car and air travel as much as you like you could offset the carbon emissions (see page 7). Many organisations have set up their own carbon offset schemes and use the money collected for local environmental projects. Visit www.climatecare.org or do a Google search for carbon offsetting. The government is trying to set up a code or standard for offsetting.

4. Have a 'green' family day out.

Explore your local area, take a picnic to the park, walk or cycle to a local museum or cinema, or try exploring on public transport. Young children love travelling by train and bus, especially if it's something they don't normally do.

The GREEN MAN Explains...

Travel and transport in itself isn't bad for the planet (although to listen to some politicians you might think differently!) The main problem is that today's transport, especially cars and planes, use fossil fuels, mainly oil, and emit greenhouse gases.

In addition conventional engines and fuels produce polluting chemicals which are bad for our health. And roads, car parks, etc. take up space which could be used for other things.

There are various moves now to use alternative fuels. Biofuels are produced from plants. They are renewable sources of energy as long as we replace them as quickly as we use them. Sugar cane and sugar beet can both be converted into fuel. In the USA corn is being grown to produce ethanol, which can be mixed with petrol. The carbon-dioxide emissions produced when the plants are burned can be soaked up by the new plants. But it will take a lot of land to produce enough biofuels. The European Union estimates that it would take 72% of its arable land to produce enough biofuels for just 10% of its fuel use. And there are complex issues to do with using land for growing fuels rather than food, and using fossil fuels to process the raw materials into biofuels. Some scientists estimate that there will be very little reduction in greenhouse gas emissions and much damage to the environment due to mono-cropping of fuel crops. Find out more about biofuels at www.carbonfootprint.com

Electric cars produce no emissions on the road and cost less to run and maintain, but they do cause emissions at the power station. To refuel an electric car you plug into the domestic supply, and some cities have recharging places. Any 13 amp socket will do. The best alternative would be electric cars refuelled at solar-power stations, but no one has built one yet!

A hybrid car has a petrol engine and an electric motor and you don't have to recharge it. The batteries are charged when the car brakes or cruises. Fuel consumption is about 30% less than for the same size petrol car, but hybrid cars are quite expensive to buy.

Air transport

Travelling by plane is bad for the planet. In a few hours flight we burn up hundreds of litres of oil and produce tonnes of CO_2 which is the main greenhouse gas causing global warming, along with water vapour, and other gases such as nitrogen oxide. Scientists estimate that the effect of these greenhouse gases on the climate is much greater when they are produced high above the earth. Some estimates suggest that at high altitudes the effects of these gases is nearly three times higher.

Aviation is responsible for between only 4 and 9% CO_2 emissions globally at present – but the number of flights is growing very fast. In the UK carbon emissions from aviation will be four to ten times greater in 2050 than in 1990 and this is likely to cancel out the effect of the reductions made by all other businesses.

Shocking statistic

- Less than 50% of the UK population fly at all. 25% take one trip a year. The top 10% of earners fly the most and the bottom 10% hardly ever.

Fuel station of the future?

Photovoltaic cells produce electricity direct from sunlight. At the moment this process is not seen as cost effective, so no-one has yet built a solar powered fuel station for electric cars.

home sweet Home

We can reduce the impact we have on the planet by making changes in our own small environment – our house and garden. Most of us spend at least 12 to 15 hours in our homes each day, more at the weekend. So the quality of the environment is important. By making greener choices here we can improve our own and our children's health and have a healthy environment to live in. By not using synthetic materials and chemicals made from fossil fuels we can help the planet.

NASTY CHEMICALS? WE'LL SORT 'EM OUT!!

A healthy house

Our homes are often polluted. Many children suffer from asthma and allergies.

- Clean less thoroughly and less often, using natural cleaners such as lemon and vinegar – experts now think that a little dirt helps the immune system! But wash your hands often with soap to stop germs spreading.

- Open windows and get rid of condensation, use extractor fans in bathrooms and kitchens.
- Plants help purify the air by absorbing chemicals – try spider plants, palms, ferns, rubber plants and ivy – see www.flowers.org.uk
- Buy furnishings which do not emit chemicals such as formaldehyde (see The Green Man Explains page 29).
- Buy natural paints and cleaners such as borax or soda crystals.
- Get your boiler checked to make sure no carbon monoxide is being given off.
- Check the labels of everything you buy and avoid anything which says 'could be hazardous to health' or lists chemicals you don't understand.
- Light a match in a room to get rid of bad smells – aerosol air fresheners have been shown to affect babies' health.
- Buy natural soaps, washing powders and toiletries – read the labels.

No-cost or low-cost savers

- Buy white recycled toilet paper. The colour dye adds unnecessary chemicals to the environment.
- Switch to natural (home-made) or ecological cleaners – they make it easier for the sewage system. You can buy some ecological cleaners, such as Ecover products, in supermarkets, health food shops or on the internet.
- Don't put toxic household wastes, such as paints, down the drain. Take them to your local recycling centre.
- Avoid using aluminium foil and cling film whenever possible. If you do use it, wash and recycle it.
- Compost paper towels if you use them.
- Bicarbonate of soda with water in a paste can be used to clean ovens and cooker tops.
- Do a 'half and half' test to compare natural cleaners with bought chemical cleaners. Can you tell the difference?

Activities

1. Experiment with natural cleaners

- Vinegar is a natural cleaner, disinfectant and deodoriser. It can be used, diluted in hot water, on hard surfaces, such as baths, floors, worktops and windows. Wipe thoroughly. Glass will shine after being washed with vinegar and water. Or try using newspapers to clean windows. Vinegar or lemon can also be used instead of bleach to remove lime scale from the toilet.
- Cut a lemon in half, sprinkle baking soda on it, and use it to scrub dishes, pots and pans, surfaces with dirt and stains. Make a paste of lemon juice and baking soda, and use it to get rid of soap scum and lime scale around taps. Shine metal, such as brass and copper, with lemon juice on a cloth.
- Mix half a cup of lemon juice with half a cup of olive oil and use it to polish wooden surfaces.
- Baking soda is a deodoriser. Use it diluted in warm water to wash out fridges and freezers.

2. Make traps for slugs and snails

Put old grapefruit skin halves up-side down near precious plants – check them daily and dispose of the slugs and snails you find inside. Or put a small amount of beer or milk in an old plastic pot and put it into the ground – the slugs will drown. Or keep slugs away by sprinkling crushed egg shells or coffee grounds around favourite plants.

3. Make your own seed pots ...

and grow some food. Re-use food pots, such as yoghurt pots, or use old egg boxes, or make newspaper pots for growing seedlings in. Newspaper pots can be put straight into the soil and decompose. They are especially good for growing carrots, parsnips and parsley, whose roots hate being disturbed.

For seed pots from old newspaper: just roll the newspaper round a bottle and tape it into a tube shape. Slide the bottle about 50mm inside the tube. Push the ends over and tape it closed to form the pot base. Take the bottle out. Cut the pot from the rest of the tube, about 100mm deep. If enough tube remains, repeat the process to make another pot.

Shocking statistics

- The newer your home and furnishings the more likely it has indoor pollution.
- More than 70,000 different chemicals are used in household cleaners and toiletries, but fewer than 1 in 4 have been tested for safety. Some are tested on animals.
- 24 million tonnes of green garden waste are sent to landfill each year in the UK – enough to fill more than a million lorries.

The challenge!

How planet friendly can you make your home?

Choose one room and make a list of the things in it, furniture, etc. and the cleaners you use in it. Which things are made from natural materials? Which may be giving off harmful chemicals? Can you change the products you use?

Do the same for other rooms and in the garden.

Garden savers

- Compost your kitchen waste and weeds – don't take garden refuse to the landfill. (See page 11).
- Garden organically – by not using pesticides and fertilisers made from fossil fuels you protect your health and help the planet. Remove pests by other means, or live with them! Visit www.gardenorganic.org.uk
- Avoid bonfires. Larger prunings can be taken to the local municipal composting site or if you have a really big garden you may need to hire a shredder a couple of times a year.
- Don't use a patio heater – put a jumper or coat on instead.
- Don't mow the lawn too short – it will need less water and provide a better habitat for small creatures.
- Use mulch around plants to conserve water and reduce weeds.
- Use peat-free compost – peat is a fossil fuel which holds carbon.
- Encourage the pest eaters – birds, hedgehogs and toads by providing food and habitats they like. Visit www.wildlifetrust.org

The GREEN MAN Explains...

Household pollution

The most common reasons for indoor air pollution are smoking, dust mites, animal hairs and toxic chemicals from building materials and furnishings, from paints, toiletries, and cleaning products. Many modern cleaners contain strong chemicals which affect us. Children are particularly at risk from chemicals used in cleaning. For example, washing powder can contain phosphates, enzymes, chlorine, formaldehyde, pesticide residues, foaming agents and synthetic colours and perfumes. Visit www.panda.org or www.Greenpeace.org to find out more.

Nasty chemicals called VOCs (volatile organic compounds) are given off by many of the ordinary things we have at home – furnishings, carpets and plastics. Curtains and furniture emit toxins such as styrene and formaldehyde, especially when new. Read the labels carefully and ask before you buy.

Plastics and synthetic materials contain chemicals which can affect our bodies – we may not notice any difference in the short term but they can have long-term effects on our health, affecting our hormones and endocrine systems, even our fertility. Visit www.foe.co.uk or www.greenpeace.org.uk to find out more.

And many of these synthetic products are manufactured using ingredients from petrochemicals – using natural products is better for the planet too.

Visit www.ecover.com or www.thegreenshop for natural cleaning products which are better for your health and for the environment.

cool your Shopping

Let's face it - we're a consumer society and that's the way the retailers like it. But it's not a sustainable way of living – we're using and wasting far too many resources. So how can we use less?

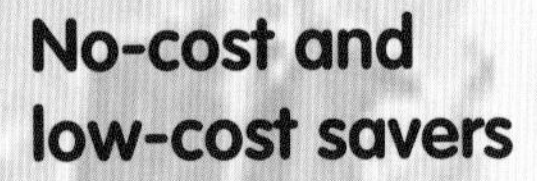

No-cost and low-cost savers

- Stop before you buy. Wait a while – often the need to have it goes away.
- Visit the shops less – go once a month instead of every week. On the other days do something else you enjoy.
- Go shopping with a list and stick to it.
- Avoid buying goods with lots of packaging.
- Share or hire items you don't use very often, such as party glasses or DIY tools.
- Buy second-hand, from charity shops, car boot sales and e-bay – this needn't be a poor-quality option. Second-hand furniture is often far better quality than new. And it's great fun finding bargains!

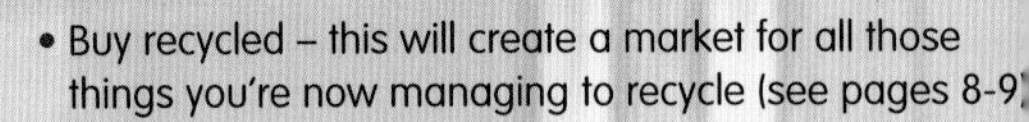

- Buy recycled – this will create a market for all those things you're now managing to recycle (see pages 8-9)
- Buy recyclable – buy things like washing liquids in recyclable containers and get them refilled.
- Buy things which last – high quality goods often last longer.
- Buy non-disposable – cloth nappies and kitchen cloths rather than paper towels. They're washable, use fewer resources and don't add to landfill. If you must have paper, buy recycled ones and compost them after use.
- Buy online or shop locally – it saves time and reduces your energy consumption.
- Buy Fairtrade and be sure that small producers in the developing world are getting a fair price or fair wage.
- Buy from responsible retailers with clear, green and ethical policies.
- Use your purchasing power to influence companies – see www.ethicalconsumer.org or www.greenchoices.org

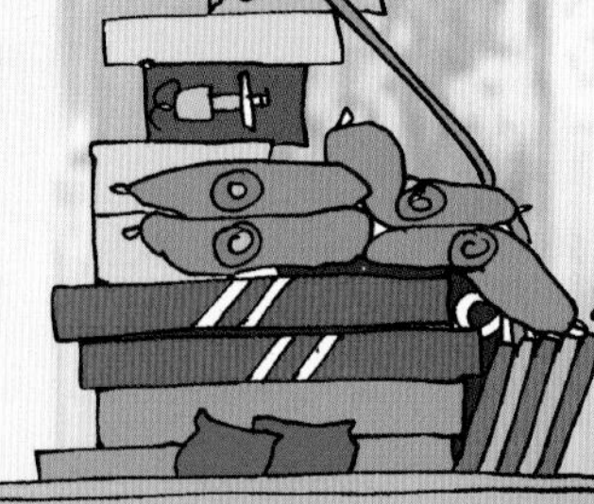

Be buy-shy

Before you buy something new ask yourself these questions:

- Do I NEED this?
- Can I borrow or hire one?
- Can I get it second-hand?
- Do I already have something I can use instead?
- What will happen if I don't buy this?

Activities

1. Check your shopping habits

How often do you go shopping? Make a list of purchases n the last week, apart from food. Highlight those items that were really necessary. What could you have managed quite happily without? What could you have borrowed or hired? How much money could you have saved?

2. Family fun fund

Save money by not buying things and set up a family un fund. Use it to pay for planet-friendly treats: fun activities, new experiences or learning new things. Decide how it's spent by voting or have one person choose each month or each 6 months. Things such as earning a sport or skill, going out for a family meal or to he cinema are mostly much better for the planet than buying products, because they use human resources rather than natural, non-renewable ones.

3. What's in a weekend?

What do you do over a typical weekend? How could you make your weekends more sustainable and have more un or save money? Could you do food shopping at a armer's market, or online? Would you enjoy playing sport or cycling in the countryside? Plan a sustainable amily weekend.

4. Gift tokens

We all get gifts we're not sure what to do with. Instead of buying gifts for birthdays or Christmas why not give your me or make something? What talents and skills do you nd your family have? What could you offer each other nd other people? Offers could include baby-sitting, omputer skills, DIY, a lift somewhere, a dream meal. ou could make a suitable card and put your 'gift oken' inside.

ome communities use this form of exchange hrough Local Exchange Trading Systems (LETS) find out more at www.gmlets.u-et.com/explore/found/intro.html

The challenge:

Resist buying! Reduce the amount you buy. Can you save £50 a month as a family by buying less?

The GREEN MAN Explains...

Why does shopping matter?

Everything we buy uses resources to make. Some of these may be renewable resources, such as wood, but it can be difficult to find out what kind of forest has been cut down to make the object, and whether environmental damage has been caused. (Look for the Forest Stewardship Council logo.) Most of the things we buy have also used non-renewable resources, such as fossil fuels and minerals. Plastics, for example, are made from oil by-products. Fossil fuels have been used to manufacture and transport them and so they contribute to global warming.

There has been a huge increase in the amount of shopping in the last twenty years and shopping is a major leisure activity. This means that we are using more of the world's resources. In the past most people had to save up for things, and they couldn't buy everything they wanted, which meant that they had time to think about whether they really needed something. By thinking about things differently and changing our habits a little we can help the planet and also save ourselves money.

The other dimension to shopping is avoiding exploitation of the people producing the goods. Huge quantities of cheap goods are now imported from China, India and other parts of the world. Sometimes these cheap prices are achieved by people working long hours (sometimes over 80 hours a week) for low wages, in unsafe conditions. Even companies which we usually trust find it difficult to monitor their suppliers and it isn't easy to find out how the product you buy has been produced.

Visit www.getethical.com to find out more.

Exciting inspiration

web-links and books – how to find out more

General

The Centre for Alternative Technology
www.cat.org.uk - ideas, information, books and technology
Friends of the Earth
www.foe.co.uk
Greenpeace
www.greenpeace.org – campaigns on environmental issues
www.greenchoices.org
simple direct information on how to make good choices
www.climatechallenge.gov.uk
new government website
www.tyndall.ac.uk
The Tyndall Centre for Climate Change Research
www.cru.uea.ac.uk
University of East Anglia's Institute of Climate Change
www.ukcip.org.uk
UK Climate Change Programme
www.bbc.co.uk/climate
interesting information, games, etc
www.doingyourbit.org.uk
DEFRA and DTLR government site to encourage everyone to take action
www.direct.gov.uk/en/Environmentandgr eenerliving/index.htm
lots more good ideas
www.afsl.org.uk
Action for Sustainable Living
www.wen.org.uk
Women's Environmental Network
www.envirolink.org
links to many environmental organisations
How we can save the planet
Hillan, M & Fawcett T. Penguin 2004
No Waste like Home
Penny Poyzer. Virgin Books Ltd 2005
Save cash and save the planet
Smith, A & Baird, N. Harper Collins 2005

Carbon emissions

See www.nef.org.uk or www.resurgence.org
For information about renewable energy companies see
www.carbonfootprint.com
www.co2balance.com
www.bestfootforward.com
www.carbonplanet.com
www.carbonneutral.com
www.foe.co.uk
www.CRed.uk.org
for carbon calculators and other information
www.climatecare.org
www.thec-changetruct.com
about off-setting

Waste and recycling

www.wasteonline.org.uk
waste at home
www.wastewatch.org.uk
www.defra.gov.uk/environment/waste
www.environment-agency.gov.uk
www.direct.gov.uk/greenerliving
information on recycling
www.recyclemore.com
www.freecycle.org
for free local recycling
Composting
see www.gardenorganic.org.uk/organicgardening/compost
www.envocare.co.uk/makingcompost
www.compost.org.uk
All about Composting
Pauline Pears HDRA

Energy

www.carbontrust.co.uk
www.est.org.uk
www.energysavingtrust.org.uk/commit
Government white paper 'The Energy Challenge'
http://www.dti.gov.uk/energy/review/page31995.html
www.greenelectricity.org
www.good-energy.co.uk
100% of its energy from renewables
www.ecotricity.co.uk
electricity from the grid but invests in renewables
www.greenenergy.uk.com
79% of its energy from renewables

Water

www.bristol-water.co.uk/environment/waterSavingAudit.asp
good water audit
www.wateraid.org.uk/uk/learn_zone/games
a global perspective and some fun games
www.water.org.uk

Food

www.food.gov.uk
food standards agency
www.vegboxschemes.co.uk
www.farmersmarkets.net
www.fairtrade.co.uk
www.soilassociation.org.uk
www.localfoodworks.org
www.foe.co.uk/campaigns/real_food/
Friends of the Earth site – looking at a range of issues relating to food and see also www.gardenorganic.org.uk
www.permaculture.org.uk
New Complete Self-Sufficiency: The classic guide for realists and dreamers
John Seymour, Dorling Kindersley
Food for free
Richard Mabey Collins 2007
Not on the label: What really goes into the food on your plate,
Lawrence, F. Penguin 2004

Transport

Environmental Transport Association
www.eta.co.uk
Transport 2000
www.transport2000.org.uk – promotes sustainable transport
car-share clubs
see www.car-plus.org.uk
'walking buses'
www.travelwise.org.uk
biofuels
www.carbonfootprint.com
www.Sustrans.org.uk
UK's leading sustainable transport charity
www.vcarfueldata.org.uk
to find out about your car's carbon emissions
www.travelcalculator.org
to work out your annual travel emissions
www.vca.gov.uk
gives CO_2 emissions of new cars
www.cyclenetwork.org.uk
UK cycle campaign groups
www.learntoletgo.uk
how to be car-free

House and garden

www.gardenorganic.org.uk
www.wildlifetrust.org
www.panda.org
www.Greenpeace.org
www.foe.co.uk
www.greenpeace.org.uk
to find out more.
www.ecover.com
www.thegreenshop
Organic Gardening
Lawrence D Hills, Penguin

Shopping and swapping

www.ethicalconsumer.org
www.greenchoices.org
www.getethical.com
Local Exchange Trading Systems (LETS)
www.letslink.org and www.gmlets.unet.com/explore/found/intro.html and www.timebanks.co.
www.freecycle.org and www.gumtree.com and www.free2collect
for free items
www.readitswapit.co.uk/TheLibrary.asp
sign up to swap/recycle books

Note website and other details are correct at tim of going to print only.